Scholastic
Vocabulary

The Moon

Lydia Carlin

SCHOLASTIC INC.

NEW YORK • TORONTO • LONDON • AUCKLAND • SYDNEY
MEXICO CITY • NEW DELHI • HONG KONG • BUENOS AIRES

ISBN-13: 978-0-545-00732-0 / ISBN-10: 0-545-00732-1

Photos Credits:
Cover: © Phototake/Alamy; title page: © Imagestate/Alamy; contents page, from top: © Jeff Taflan/zefa/Corbis, © Marvin E. Newman/Getty Images, © NASA Handout/Getty Images; page 4: © Jeff Taflan/zefa/Corbis; page 5: © Phototake/Alamy, inset: © Lefty's Editorial Services/Jim McMahon; page 6: © StockTrek/Getty Images; page 7, from top: © Digital Vision/Getty Images, © NASA/Apollo/Digital Version by Science Faction; page 8: © David Frazier/Corbis; page 9: © Patrick Eden/Alamy; page 10, all: © Larry Landolfi/Photo Researchers; page 11: © Marvin E. Newman/Getty Images, inset © Detlev Ravenswaay/Photo Researchers; page 12: © Bettmann/Corbis; page 13: © NASA Handout/Getty Images; page 14: © Roger Ressmeyer/Getty Images; page 15: © Richard T. Nowitz/Corbis; page 16: © Larry Landolfi/Photo Researchers; back cover: © Brand X Pictures/Alamy.

Photo research by Dwayne Howard
Design by Holly Grundon

12 11 10 9   40   10 11 12/0

Printed in the U.S.A.
First printing, December 2007

Contents

Chapter 1

What Is the Moon?

Some stories say the moon is a smiling face. Others say it is made of green cheese. But what is the moon really? Keep reading to find out.

Fast Fact

Earth

moon

It takes about 27 days for the moon to go around the Earth one time.

The moon is a **satellite**. Satellites travel around planets. The moon circles our planet, the Earth.

Fast Fact

The Earth is about four times bigger than the moon.

The moon is round and hard. It is a lonely, silent place. Days are broiling hot. Nights are freezing cold.

moon crater

moon rock

The moon is covered with deep holes called **craters**. It has huge rocks. It even has mountains!

Chapter 2

The Moon and the Earth

The moon is the Earth's closest neighbor in space. That is why the moon looks so big in the night sky.

Fast Fact

The moon's gravity pulls on our oceans, causing water to rise and fall.

The moon can seem very bright. But it does not make its own light. What you see is the sun's light shining on the moon.

new moon

crescent moon

quarter moon

full moon

We see different parts of the moon's sunlit surface as it orbits the Earth. Sometimes we see a crescent moon. Sometimes we see a full moon. These shapes are called **phases**.

solar eclipse

Fast Fact

People protect their eyes with special sunglasses when viewing a solar eclipse.

Once in a while, the moon moves directly between the sun and the Earth. This blocks most of the sunlight from reaching the Earth. It is called a **solar eclipse**.

Chapter 3

To the Moon!

The moon is the only place in space that people have visited. How did they get there? Rocket ships!

Fast Fact

This astronaut's backpack is full of air so he can breathe.

The first **astronauts** walked on the moon in 1969. They learned that nothing lives there. There is not even any air.

Only 12 people have walked on the moon so far.

footprints on the moon

The astronauts left footprints behind when they visited the moon. The footprints will stay there for thousands of years because the moon has no wind.

Fast Fact

Kids can go to Space Camp® to learn how to be astronauts.

Would you like to go to the moon? Maybe you can train to be an astronaut. Then someday you might be able to leave your footprints there, too!

Glossary

astronaut (**ass**-truh-nawt): someone who travels in space

crater (**kray**-tur): a large hole in the ground

gravity (**grav**-uh-tee): the force that pulls things to the surface of the Earth or a heavenly body.

phase (**faze**): a shape that the moon takes as viewed from Earth

solar eclipse (**soh**-lur i-**klips**): when the moon comes between the sun and the Earth so that all or part of the sun's light is blocked out

satellite (**sat**-uh-lite): a moon or heavenly body that travels in an orbit around a larger heavenly body; a spacecraft that is sent into orbit around the Earth

Comprehension Questions

1. Can you share two ways that the moon is different from the Earth?

2. Can you share two facts about space travel to the moon?

3. Can you think of four marvelous words to describe the moon?